The Spontaneous Generation

Tyndall John

The Spontaneous Generation

LM Publishers

Part I

Within ten minutes' walk of a little cottage which I have recently built in the Alps, there is a small lake, fed by the melted snows of the upper mountains. During the early weeks of summer no trace of life is to be discerned in this water; but invariably toward the end of July, or the beginning of August, swarms of tailed organisms are seen enjoying the sun's warmth along the shallow margins of the lake, and rushing with audible patter into the deeper water at the approach of danger. The origin of this periodic crowd of living things is by no means obvious. For years I had never noticed in the lake either an adult frog, or the smallest fragment of frog spawn; so that, were I not otherwise informed, I should have found the conclusion of Mathiole a natural one, namely, that tadpoles are generated in lake-mud by the vivifying action of the sun.

The checks which experience alone can furnish being absent, the spontaneous generation of creatures quite as high as the frog in the scale of being was assumed for ages to be a fact. Here, as elsewhere, the dominant mind of Aristotle stamped its notions on the world at large. For nearly twenty centuries after him men found no difficulty in believing in cases of spontaneous generation which would now be rejected as monstrous by the most fanatical supporter of the doctrine. Shell-fish of all kinds were considered to be without parental origin. Eels were supposed to spring spontaneously from the fat ooze of the Nile. Caterpillars were the spontaneous products of the leaves on which they fed, while winged insects, serpents, rats, and mice, were all thought capable of being generated without sexual intervention.

The most copious source of this life without an ancestry was putrefying flesh, and, lacking the checks imposed by fuller investigation, the conclusion that flesh possesses and exerts this generative power is a natural one. I well remember when a child of ten or twelve seeing a joint of imperfectly salted beef cut into, and coils of maggots laid bare within the mass. Without a moment's hesitation I jumped to the conclusion that these maggots had been spontaneously generated in the meat. I had no knowledge which could qualify or oppose this conclusion, and for the time it was irresistible. The childhood of the individual typifies that of the race, and the belief here enunciated was that of the world for nearly two thousand years.

To the examination of this very point the celebrated Francesco Redi, physician to the Grand-dukes Ferdinand II. and Cosmo III. of Tuscany, and a member of the Academy del

Cimento, addressed himself in 1668. He had seen the maggots of putrefying flesh, and reflected on their possible origin. But he was not content with mere reflection, nor with the theoretic guess-work which his predecessors had founded upon their imperfect observations. Watching meat during its passage from freshness to decay, prior to the appearance of maggots he invariably observed flies buzzing round the meat and frequently alighting on it. The maggots, he thought, might be the half-developed progeny of these flies.

The inductive guess precedes experiment, by which, however, it must be finally tested. Redi knew this, and acted accordingly. Placing fresh meat in a jar and covering the mouth with paper, he found that though the meat putrefied in the ordinary way, it never bred maggots, while the same meat placed in open jars soon swarmed with these organisms. For the paper cover he then substituted fine gauze, through which the odor of the meat could rise. Over it the flies buzzed, and on it they laid their eggs, but, the meshes being too small to permit the eggs to fall through, no maggots were generated in the meat. They were, on the contrary, hatched upon the gauze. By a series of such experiments Redi destroyed the belief in the spontaneous generation of maggots in meat, and with it doubtless many related beliefs. The combat was continued by Vallisneri, Schwammerdam, and Réaumur, who succeeded in banishing the notion of spontaneous generation from the scientific minds of their day. Indeed, as regards such complex organisms as those which formed the subject of their researches, the notion was banished forever.

But the discovery and improvement of the microscope, though giving a death-blow to much that had been previously written and believed regarding spontaneous generation, brought also into view a world of life formed of individuals so minute—so close as it seemed to the ultimate particles of matter—as to suggest an easy passage from atoms to organisms. Animal and vegetable infusions exposed to the air were found clouded and crowded with creatures far beyond the reach of unaided vision, but perfectly visible to an eye strengthened by the microscope. With reference to their origin these organisms were called "infusoria." Stagnant pools were found full of them, and the obvious difficulty of assigning a germinal origin to existences so minute furnished the precise condition necessary to give new play to the notion of heterogenesis or spontaneous generation.

The scientific world was soon divided into two hostile camps, the leaders of which alone can here be briefly alluded to. On the one side we have Buffon and Needham, the former postulating his "organic molecules," and the latter assuming the existence of a special "vegetative force" which drew the molecules together so as to form living things. On the other side we have the celebrated Abbé Lazzaro Spallanzani, who in 1877 published results counter to those announced by Needham in 1748, and obtained by methods so precise as to completely overthrow the convictions based upon the labors of his predecessor. Charging his flasks with organic infusions, he sealed their necks with the blow-pipe, subjected them in this condition to the heat of boiling water, and subsequently exposed them to temperatures favorable to the development of life. The infusions continued unchanged for months, and when the flasks were subsequently opened no trace of life was found.

Here I may forestall matters so far as to say that the success of Spallanzani's experiments depended wholly on the locality in which he worked. The air around him must have been free from the more obdurate infusorial germs, for otherwise the process he followed would, as was long afterward proved by Wyman, have infallibly yielded life. But his refutation of the doctrine of spontaneous generation is not the less valid on this account. Nor is it in any way upset by the fact that others in repeating his experiments obtained life when he obtained none. Rather is the refutation strengthened by such differences. Given two experimenters equally skillful and equally careful, operating in different places on the same infusions, in the same way, and assuming the one to obtain life while the other fails to obtain it; then its well-established absence in the one case proves that some ingredient foreign to the infusion must be its cause in the other.

Spallanzani's sealed flasks contained but small quantities of air, and as oxygen was afterward shown to be generally essential to life, it was thought that the. absence of life observed by Spallanzani might have been due to the lack of this vitalizing gas. To dissipate this doubt, Schulze in 1836 half-tilled a flask with distilled water, to which animal and vegetable matters were added. First boiling his infusion to destroy whatever life it might contain, Schulze sucked daily into his flask air which had passed through a series of bulbs containing concentrated sulphuric acid, where all germs of life suspended in the air were supposed to be destroyed. From May to August this process was continued without any development of infusorial life.

Here, again, the success of Schulze was due to his working in comparatively pure air, but even in such air his experiment is a very risky one. Germs will pass, unwetted and unscathed, through sulphuric acid, unless the most special care is taken to detain them. I have repeatedly failed, by repeating Schulze's experiments, to obtain his results. Others have failed likewise. The air passes in bubbles through the bulbs, and, to render the method secure, the passage of the air must be so slow as to cause the whole of its floating matter, even to the core of each bubble, to touch the surrounding liquid. But, if this precaution be observed, water will be found quite as effectual as sulphuric acid. By the aid of an air pump, in a highly-infective atmosphere, I have thus drawn air for weeks without intermission, first through bulbs containing water, and afterward through vessels containing organic infusions, without any appearance of life. The germs were not killed, but they were effectually intercepted, while the objection that the air had been injured by being brought into contact with strongly corrosive substances was avoided.

The brief paper of Schulze, published in Poggendorf's *Annalen* for 1836, was followed in 1837 by another short and pregnant communication by Schwann. Redi, as we have seen, traced the maggots of putrefying flesh to the eggs of flies. But he did not and he could not know the meaning of putrefaction itself. He had not the instrumental means to inform him that *it* also is a phenomenon attendant on the development of life. This was first proved in the paper now alluded to. Schwann placed flesh in a flask filled to one-third of its capacity with water, sterilized the flask by boiling, and then supplied it for months with calcined air. Throughout this time "there appeared no mould, no infusoria, no putrefaction; the flesh remained unaltered, while the liquid continued as clear as it was immediately after boiling." Schwann then varied his experimental argument, with no alteration in the result. His final conclusion was, that putrefaction is due to decompositions of organic matter attendant on the multiplication therein of minute organisms. These organisms were derived not from the air, but from something contained in the air, which was destroyed by a sufficiently high temperature. There never was a more determined opponent of the doctrine of spontaneous generation than Schwann, though a strange attempt was made a year and a half ago to enlist him and others equally opposed to it on the side of the doctrine.

The physical character of the agent which produces putrefaction was further revealed by Helmholtz in 1843. By means of a membrane he separated a sterilized putrescible liquid from a putrefying one. The sterilized infusion remained perfectly intact. Hence it was not the liquid of the putrefying mass—for it could freely diffuse through the membrane—but something contained in the liquid, and which was stopped by the membrane, that caused the putrefaction. In 1854 Schroeder and Von Dusch struck into this inquiry, which was subsequently followed up by Schroeder alone. These able experimenters employed plugs of cotton-wool to filter the air supplied to their infusions. Fed with such air, in the great majority of cases the putrescible liquids remained perfectly sweet after boiling. Milk formed a conspicuous exception to the general rule. It putrefied after boiling, though supplied with carefully-filtered air. The researches of Schroeder bring us up to the year 1859.

In that year a book was published which seemed to overturn some of the best-established facts of previous investigators. Its title was "Hétérogénie," and its author was F. A. Pouchet, Director of the Museum of Natural History at Rouen. Ardent, laborious, learned, full not only of scientific but of metaphysical fervor, he threw his whole energy into the inquiry. Never did a subject require the exercise of the cold, critical faculty more than this one— calm study in the unraveling of complex phenomena, care in the preparation of experiments, care in their execution, skillful variation of conditions, and incessant questioning of results, until repetition had placed them beyond doubt or question. To a man of Pouchet's temperament, the subject was full of danger—danger not lessened by the theoretic bias with which he approached it. This is revealed by the opening words of his preface:

"Lorsque, par la méditation, il fut évident pour moi que la génération spontanée était encore l'un des moyens qu'emploie la nature pour la reproduction des êtres, je m'appliquai a découvrir par quels procédés on pouvait parvenir à en mettre les phénomènes en évidence." It is needless to say that such a prepossession required a strong curb. Pouchet repeated the experiments of Schulze and Schwann with results diametrically opposed to theirs. He heaped experiment upon experiment, and argument upon argument, spicing with the sarcasm of the advocate the logic of the man of science. In view of the multitudes required to produce the observed results, be ridiculed the assumption of atmospheric germs. This was one of his strongest points. "Si les Proto-organismes que nous voyons pulluler partout et dans tout, avaient leurs germes disséminés dans l'atmosphère, dans la proportion mathématiquement indispensable à cet effet, l'air en serait totalement obscurci, car ils devraient s'y trouver beaucoup plus serres que les globules d'eau qui forment nos nuages épais. II n'y a pas là la moindre exagération." Recurring to the subject, he exclaims, "L'air dans lequel nous vivons aurait presque la densité du fer." There is often a virulent contagion in a confident tone, and this hardihood of argumentative assertion was sure to influence minds swayed not by knowledge, but by authority. Had Pouchet known that "the blue ethereal sky" is formed of suspended particles, through which the sun freely shines, he would hardly have ventured upon this line of argument.

Pouchet's pursuit of this inquiry strengthened the conviction with which he began it, and landed him in downright credulity in the end. I do not question his ability as an observer, but the inquiry needed a disciplined experimenter. This latter implies not mere ability to look at things as Nature offers them to our inspection, but to force her to show herself under conditions prescribed by the experimenter himself. Here Pouchet lacked the necessary discipline. Yet the vigor of his onset raised clouds of doubt, which for a time obscured the whole field of inquiry. So difficult indeed did the subject seem, and so incapable of definite solution, that when Pasteur made known his intention to take it up, his friends Biot and Dumas expressed their regret, earnestly exhorting him to set a definite and rigid limit to the time he purposed spending in this apparently unprofitable field.

Schooled by his education as a chemist, and by special researches on the closely related question of fermentation, Pasteur took up this subject under particularly favorable conditions.

His work and his culture had given strength and finish to his natural aptitudes. In 1862, accordingly, he published a paper "On the Organized Corpuscles existing in the Atmosphere," which must forever remain classical. By the most ingenious devices he collected the floating particles of the air surrounding his laboratory in the Rue d'Ulm, and subjected them to microscopic examination. Many of them he found to be organized particles. Sowing them in sterilized infusions, he obtained abundant crops of microscopic organisms. By more refined methods he repeated and confirmed the experiments of Schwann, which had been contested by Pouchet, Montegazza, Joly, and Musset. He also confirmed the experiments of Schroeder and Von Dusch. He showed that the cause which communicated life to his infusions was not uniformly diffused through the air; that there were aerial interspaces which possessed no power to generate life. Standing on the Mer de Glace, near the Montanvert, he snipped off the ends of a number of hermetically-scaled flasks containing organic infusions. One out of twenty of the flasks thus supplied with glacier air showed signs of life afterward, while eight out of twenty of the same infusions, supplied with the air of the plains, became crowded with life. He took his flasks into the caves under the Observatory of Paris, and found the still air in these caves devoid of generative power. These and other experiments, carried out with a severity perfectly obvious to the instructed scientific reader, and accompanied by a logic equally severe, restored the conviction that, even in these lower reaches of the scale of being, life does not appear without the operation of antecedent life.

The main position of Pasteur, though often assailed, has never yet been shaken. It has, on the contrary, been strengthened by practical researches of the most momentous kind. He has applied the knowledge won from his inquiries to the preservation of wine and beer, to the manufacture of vinegar, to the staying of the plague which threatened utter destruction to the silk-husbandry of France, and to the examination of other formidable diseases which assail the higher animals, including man. His relation to the improvements which Prof. Lister has introduced into surgery is shown by a letter quoted in his "Études sur la Bière." Prof. Lister there expressly thanks Pasteur for having given him the only principle which could have conducted the antiseptic system to a successful issue. The strictures regarding Pasteur's defects of reasoning, to which we have been lately accustomed, delivered with a tone of supercilious contempt, where reverent teachableness would have been the fitting state of mind, throw abundant light upon their author, but none upon Pasteur.

Redi, as we have seen, proved the maggots of putrefying flesh to be derived from the eggs of flies; Schwann proved putrefaction itself to be the concomitant of far lower forms of life than those dealt with by Redi. Our knowledge here, as elsewhere in connection with this subject, has been vastly extended by Prof. Cohn, of Breslau. "No putrefaction," he says, "can occur in a nitrogenous substance if its bacteria be destroyed and new ones prevented from entering it. Putrefaction begins as soon as bacteria, even in the smallest numbers, are admitted either accidentally or purposely. It progresses in direct proportion to the multiplication of the bacteria, it is retarded when they exhibit low vitality, and is stopped by all influences which either hinder their development or kill them. All bactericidal media are therefore antiseptic and disinfecting." It was these organisms acting in wound and abscess which so frequently converted our hospitals into charnel-houses, and it is their destruction by the antiseptic system that now renders justifiable operations which no surgeon would have attempted a few years ago. The gain is immense—to the practising surgeon as well as to the patient practised upon. Contrast the anxiety of never feeling sure whether the most brilliant operation might not be rendered nugatory by the access of a few particles of unseen hospital-dust, with the comfort derived from the knowledge that all power of mischief on the part of such dust has been surely and certainly annihilated. But the action of living contagia extends beyond the domain of the surgeon. The power of reproduction and indefinite self-multiplication which is characteristic of living things, coupled with the undeviating fact of contagia "breeding true," has given strength and consistency to a belief long entertained by penetrating minds that epidemic diseases generally are the concomitants of parasitic life. "There begins to be faintly visible to us a vast and destructive laboratory of Nature wherein the diseases which are most fatal to animal life, and the changes to which dead organic matter is passively liable, appear bound together by what must at least be called a very close analogy of causation." According to this view, which, as I have said, is daily gaining converts, a contagious disease may be defined as a conflict between the person smitten by it and a specific organism which multiplies at his expense, appropriating his air and moisture, disintegrating his tissues, or poisoning him by the decompositions incident to its growth.

During the ten years extending from 1859 to 1869, researches on radiant heat in its relations to the gaseous form of matter occupied my continual attention. When air was experimented on, I had to cleanse it effectually of floating matter, and, while doing so, I was surprised to notice that, at the ordinary rate of transfer, such matter passed freely through alkalies, acids, alcohols, and ethers. The eye being kept sensitive by darkness, a concentrated beam of light was found to be a most searching test for suspended matter both in water and in air—a test indeed indefinitely more searching and severe than that furnished by the most powerful microscope. With the aid of such a beam I examined air filtered by cotton-wood, air long kept free from agitation, so as to allow the floating matter to subside, calcined air, and air filtered by the deeper cells of the human lungs. In all cases the correspondence between my experiments and those of Schroeder, Pasteur, and Lister, in regard to spontaneous generation, was perfect. The air which they found inoperative was proved by

the luminous beam to be optically pure and therefore germless. Having worked at the subject both by experiment and reflection, on Friday evening, the 21st of January, 1870, I brought it before the members of the Royal Institution. Two or three months subsequently, for sufficient practical reasons, I ventured to direct public attention to the subject in a letter to the *Times*. Such was my first contact with this important question.

This letter, I believe, gave occasion for the first public utterance of Dr. Bastian in relation to this question. He did me the honor to inform me, as others had informed Pasteur, that the subject "pertains to the biologist and physician." He expressed "amazement" at my reasoning, and warned me that before what I had done could be undone "much irreparable mischief might be occasioned." With far less preliminary experience to guide and warn him, Dr. Bastian was far bolder than Pouchet in his experiments, and far more adventurous in his conclusions. With organic infusions he obtained the results of his celebrated predecessor, but he did far more—the atoms and molecules of inorganic liquids passing under his manipulation into those more "complex chemical compounds" which we dignify by calling them "living organisms." For five years, or thereabouts, Dr. Bastian ploughed the field without impediment from me, and, now that one can overlook the work, I am bound in truth to say that very wonderful ploughing it has been. As regards the public who take an interest in such things, and apparently also as regards a large portion of the medical profession, he certainly succeeded in restoring the subject to a state of uncertainty similar to that which followed the publication of Pouchet's volume in 1859.

It is desirable that this uncertainty should be removed from the public mind, and doubly desirable on practical grounds that it should be removed from the minds of medical men. In the present article, therefore, I propose discussing this question face to face with some eminent and fair-minded member of the medical profession who, as regards spontaneous generation, entertains views adverse to mine. Such a one it would be easy to name; but it is perhaps better to rest in the impersonal. I shall therefore simply call my proposed co-inquirer my friend. With him at my side I shall endeavor, to the best of my ability, so to conduct this discussion that he who runs may read, and that he who reads may understand.

Let us begin at the beginning. I ask my friend to step into the laboratory of the Royal Institution, where I place before him a basin of thin turnip-slices barely covered with distilled water kept at a temperature of 120° Fahr. After digesting the turnip for four or five hours we pour off the liquid, boil it, filter it, and obtain an infusion as clear as filtered drinking-water. We cool the infusion, test its specific gravity, and find it to be 1006 or higher—water being 1000. A number of small, clean, empty flasks, of the shape here shown, are before us. One of them is slightly warmed with a spirit-lamp, and its open end is then dipped into the turnip-infusion. The warmed glass is afterward chilled, the air within the flask cools, contracts, and is followed in its contraction by the infusion. Thus we get a small quantity of liquid into the flask.

We now heat this liquid carefully. Steam is produced, which issues from the open neck, carrying the air of the flask along with it. After a few seconds' ebullition, the open neck is again plunged into the infusion. The steam within the flask condenses, the liquid enters to supply its place, and in this way we fill our little flask to about four-fifths of its volume. This description is typical; we may thus fill a thousand flasks with a thousand different infusions.

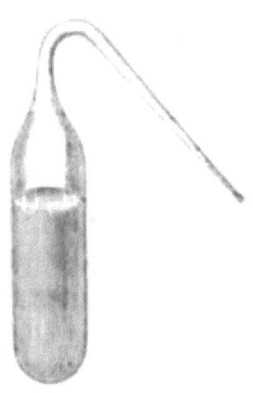

I now ask my friend to notice a trough made of sheet-copper, with two rows of handy little Bunsen burners underneath it. This trough, or bath, is nearly filled with oil; a piece of thin plank constitutes a kind of lid for the oil-bath. The wood is perforated with circular apertures wide enough to allow our small flask to pass through and plunge itself in the oil, which has been heated, say, to 250° Fahr. Clasped all round by the hot liquid, the infusion in the flask rises to its boiling-point, which is not sensibly over 212° Fahr. Steam issues from the open neck of the flask, and the boiling is continued for five minutes. With a pair of small brass tongs an assistant now seizes the neck near its junction with the flask, and partially lifts the latter out of the oil. The steam does not cease to issue, but its violence is abated. With a second pair of tongs held in one hand, the neck of the flask is seized close to its open end, while with the other hand a Bunsen's flame or an ordinary spirit-flame is brought under the middle of the neck. The glass reddens, whitens, softens, and as it is gently drawn out the neck diminishes in diameter, until the canal is completely blocked up. The tongs with the fragment of severed neck being withdrawn, the flask, with its contents diminished by evaporation, is lifted from the oil-bath perfectly sealed hermetically.

Sixty such flasks filled, boiled, and sealed, in the manner described, and containing strong infusions of beef, mutton, turnip, and cucumber, are carefully packed in saw-dust and transported to the Alps. Thither, to an elevation of about 7,000 feet above the sea, I invite my co-inquirer to accompany me. It is the month of July, and the weather is favorable to putrefaction. We open our box at the Bel-Alp, and count out fifty-four flasks, with their liquids as clear as filtered drinking-water. In six flasks, however, the infusion is found muddy. We closely examine these, and discover that every one of them has had its fragile end broken off in the transit from London. Air has entered the flasks, and the observed muddiness is the result. My colleague knows as well as I do what this means. Examined with a pocket-lens, or even with a microscope of insufficient power, nothing is seen in the muddy liquid; but regarded with a magnifying power of a thousand diameters or so, what an astonishing appearance does it present! Leeuwenhoek estimated the population of a single drop of stagnant water at 500,000,000: probably the population of a drop of our turbid infusion would be this many times multiplied. The field of the microscope is crowded with organisms, some wabbling slowly, others shooting rapidly across the microscopic field. They dart hither and thither like a rain of minute projectiles;

they pirouette and spin so quickly round, that the retention of the retinal impression transforms the little living rod into a twirling wheel. And yet the most celebrated naturalists tell us that they are vegetables. From the rod-like shape which they so frequently assume, these organisms are called bacteria—a term, be it here remarked, which covers organisms of very diverse kinds.

Has this multitudinous life been spontaneously generated in these six flasks, or is it the progeny of living germinal matter carried into the flasks by the entering air? If the infusions have a self-generative power, how are the sterility and consequent clearness of the fifty-four uninjured flasks to be accounted for? My colleague may urge—and fairly urge—that the assumption of germinal matter is by no means necessary; that the air itself may be the one thing needed to wake up the dormant infusions. We will examine this point immediately. But I would meanwhile remind my friend that I am working on the exact lines laid down by our most conspicuous heterogenist. He distinctly affirms that the withdrawal of the atmospheric pressure above the infusion favors the production of organisms; and he accounts for their absence in tins of preserved meat, fruit, and vegetables, by the hypothesis that fermentation *has* begun in such tins, that gases *have* been generated, the pressure of which has stifled the incipient life and stopped its further development. This is Dr. Bastian's theory of preserved meats. Its author has never, to my knowledge, pierced a tin of preserved meat, fruit, or vegetable, under water with a view of testing its truth. Had he done so, he would have found it erroneous. In well-preserved tins I have invariably found, not an outrush of gas, but an inrush of water, when the tin was perforated. I have noticed this recently in tins which have lain perfectly good for sixty-three years in the Royal Institution. Modern tins, subjected to the same test, yielded the same result. From time to time, moreover, during the last two years, I have placed glass tubes, containing clear infusions of turnip, hay, beef, and mutton, in iron bottles, and subjected them to air-pressures varying from ten to twenty-seven atmospheres—pressures, it is needless to say, far more than sufficient to tear a preserved meat-tin to shreds. After ten days these infusions were taken from their bottles rotten with putrefaction and teeming with life. Thus collapses an hypothesis which had no rational foundation, and which could never have seen the light had the slightest attempt been made to verify it.

Our fifty-four vacuous and pellucid flasks also declare against this heterogenist. We expose them to a warm Alpine sun by day, and at night we suspend them in a warm kitchen. Four of them have been accidentally broken; but at the end of a month we find the fifty remaining ones as clear as at the commencement. There is no sign of putrefaction or of life in any of them. We divide these flasks into two groups of twenty-three and twenty-seven respectively (an accident of counting rendered the division uneven). The question now is whether the admission of air can liberate any generative energy in the infusions. Our next experiment will answer this question and something more. We carry the flasks to a hay-loft, and there, with a pair of steel pliers, snip off the sealed ends of the group of twenty-three. Each snipping-off is of course followed by an inrush of air. We now carry our twenty-seven flasks, our pliers, and a spirit-lamp, to a ledge overlooking the Aletsch glacier, about two hundred feet above the hay-loft, from which ledge the mountain falls almost precipitously to the northeast for about a thousand feet. A gentle wind blows toward us from the northeast—that is, across the crests and snow-fields of the Oberland mountains. We are, therefore, bathed by air which must have been for a good while out of practical contact with either animal or vegetable life. I stand carefully to leeward of the flasks, for no dust or particle from my clothes or body must be blown toward them. An assistant ignites the spirit-lamp, into the flame of which I plunge the pliers, thereby destroying all attached germs or organisms. Then I snip off the sealed end of the flask. Prior to every snipping the same process is gone through, no flask being opened without the previous cleansing of the pliers by the flame. In this way we charge our twenty-seven flasks with clean vivifying mountain-air.

We place the fifty flasks, with their necks open, over a kitchen stove, in a temperature varying from 50° to 90° Fahr., and in three days find twenty-one out of the twenty-three flasks opened on the hay-loft invaded by organisms— two only of the group remaining free from them. After three weeks' exposure to precisely the same conditions, not one of the twenty-seven flasks opened in free air had given way. No germ from the kitchen-air had ascended the narrow necks, the flasks being shaped to produce this result. They are still in the Alps, as clear, I doubt not, and as free from life as they were when sent off from London.

What is my colleague's conclusion from the experiment before us? Twenty-seven putrescible infusions, first *in vacuo,* and afterward supplied with the most invigorating air, have shown no sign of putrefaction or of life. And as to the others, I almost shrink from asking him whether the hay-loft has rendered them spontaneously generative. Is not the inference here imperative that it is not the air of the loft—which is connected through a constantly-open door with the general atmosphere—but something contained in the air, that has produced the effects observed? What is this something? A sunbeam glinting through a chink in the roof or wall, and traversing the air of the loft, would show it to be laden with suspended dust-particles. Indeed, the dust is distinctly visible in the diffused daylight. Can *it* have been the origin of the observed life? If so, are we not bound by all antecedent experience to regard these fruitful particles as the germs of the life observed?

The name of Baron Liebig has been constantly mixed up with these discussions. "We have," it is said, "his authority for assuming that dead decaying matter can produce fermentation." True, but with Liebig fermentation was by no means synonymous with *life*. It will be observed, by the careful reader of Dr. Bastian's works, that whenever their author refers to this alleged power of decaying matter, he invariably couples with it the vague term "fermentation," thus softening the shock of the hypothesis which he insinuates rather than asserts. But our present intention is to brush all vagueness aside. We therefore ask, "Does the *life* of our flasks proceed from dead particles?" If my co-inquirer should reply "Yes," then I would ask him: "What warrant does Nature offer for such an assumption? Where, amid the multitude of vital phenomena in which her operations have been clearly traced, is the slightest countenance given to the notion that the sowing of dead particles can produce a living crop?" With regard to Baron Liebig, had he studied the revelations of the microscope in relation to these questions, a mind so penetrating could never have missed the significance of the facts revealed. He, however, neglected the microscope, and fell into error—but not into error so gross as that in support of which his authority has been invoked. Were he now alive, he would, I doubt not, repudiate the use often made of his name— Liebig's view of fermentation was at least a scientific one, founded on profound conceptions of molecular instability. But this view by no means involves the notion that the planting of dead particles— "Stickstoffsplittern," as Cohn contemptuously calls them is followed by the sprouting of infusorial life.

Part II

Let us now return to London and fix our attention on the dust of *its* air. Suppose a room in which the house-maid has finished her work to be completely closed, with the exception of an aperture in a shutter through which a sunbeam enters and crosses the room. The floating dust reveals the track of the light. Let a lens be placed in the aperture to condense the beam. Its parallel rays are now converged to a cone, at the apex of which the dust is raised to almost unbroken whiteness by the intensity of its illumination. Defended from all glare, the eye is peculiarly sensitive to this scattered light. The floating dust of London rooms is organic, and may be burned without leaving visible residue. The action of a spirit-lamp flame upon the floating matter has been elsewhere thus described:

"In a cylindrical beam which strongly illuminated the dust of our laboratory, I placed an ignited spirit-lamp. Mingling with the flame, and round its rim, were seen curious wreaths of darkness resembling an intensely black smoke. On placing the flame at some distance below the beam, the same dark masses stormed upward. They were blacker than the blackest smoke ever seen issuing from the funnel of a steamer; and their resemblance to smoke was so perfect as to prompt the conclusion that the apparently pure flame of the alcohol-lamp required but a beam of sufficient intensity to reveal its clouds of liberated carbon. "But is the blackness smoke? This question presented itself in a moment, and was thus answered: A red-hot poker was placed underneath the beam; from it the black wreaths also ascended. A large hydrogen-flame, which emits no smoke, was next employed, and it also produced with augmented copiousness those whirling masses of darkness. Smoke being out of

the question, what is the darkness? It is simply that of stellar space; that is to say, blackness resulting from the absence from the track of the beam of all matter competent to scatter its light. When the flame was placed below the beam, the floating matter was destroyed in situ; and the heated air, freed from this matter, rose into the beam, jostled aside the illuminated particles, and substituted for their light the darkness due to its own perfect transparency. Nothing could more forcibly illustrate the invisibility of the agent which renders all things visible. The beam crossed, unseen, the black chasm formed by the transparent air, while, at both sides of the gap, the thick-strewed particles shone out like a luminous solid under the powerful illumination."

Supposing an infusion intrinsically barren, but readily susceptible of putrefaction when exposed to common air, to be brought into contact with this unilluminable air, what would be the result? It would never putrefy. It might, however, be urged that the air is spoiled by its violent calcination. Oxygen passed through a spirit-lamp flame is, it may be thought, no longer the oxygen suitable for the development and maintenance of life. We have an easy escape from this difficulty, which is based, however, upon the unproved assumption that the air has been affected by the flame. Let a condensed beam be sent through a large flask or bolt-head containing common air. The track of the beam is seen within the flask—the dust revealing the light, and the light revealing the dust. Cork the flask, stuff its neck with cotton-wool, or simply turn it mouth downward and leave it undisturbed for a day or two. Examined afterward with the luminous beam, no track is visible; the light passes through the flask as through a vacuum. The floating matter has abolished itself, being now attached to the interior surface of the flask. Were it our object, as it will be subsequently, to effectually detain the dirt, we might coat that surface with some sticky substance. Here, then, without "torturing" the air in any way, we have found a means of ridding it, or rather of enabling it to rid itself, of floating matter.

We have now to devise a means of testing the action of such spontaneously purified air upon putrescible infusions. Wooden chambers, or cases, are accordingly constructed having glass fronts, side-windows, and back-doors. Through the bottoms of the chambers test-tubes pass air-tight, their open ends, for about one-fifth of the length of the tubes, being within the chambers. Provision is made for a free connection through sinuous channels between the inner and the outer air. Through such channels, though open, no dust will reach the chamber. The top of each chamber is perforated by a circular hole two inches in diameter and closed air-tight by a sheet of India-rubber. This is pierced in the middle by a pin, and through the pin-hole is pushed the shank of a long pipette, ending above in a small funnel. The shank also passes through a stuffing-box of cotton-wool moistened with glycerine; so that, tightly clasped by the rubber and wool, the pipette is not likely in its motions up and down to carry any dust into the chamber. The annexed woodcut shows a chamber with six test-tubes, its side-windows to w w, its pipette p C, and its sinuous channels a b which connect the air of the chamber with the outer air.

The chamber is carefully closed and permitted to remain quiet for two or three days. Examined at the beginning by a beam sent through its windows, the air is found laden with floating matter, which in three days has wholly disappeared. To prevent its ever rising again into the chambers, the internal surface is coated with glycerine. The fresh but putrescible liquid is introduced into the six tubes in succession by means of the pipette. Permitted to remain without further precaution, every one of the tubes would putrefy and till itself with life. The liquid has been in contact with dust-laden air by which it has been infected, and the infection must be destroyed.

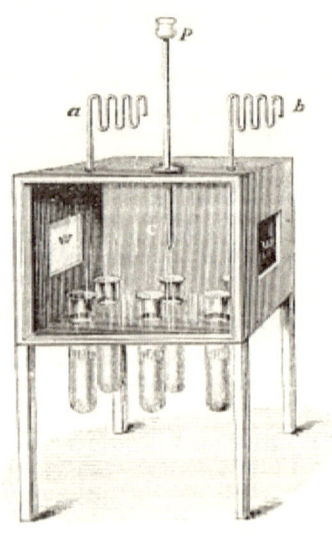

This is done by plunging the six tubes into a bath of heated oil and boiling the infusion. The time requisite to destroy the infection depends wholly upon its nature. Two minutes' boiling suffices to destroy some contagia, whereas two hundred minutes' boiling fails to destroy others. After the infusion has been sterilized, the oil-bath is withdrawn, and the liquid, whose putrescibility has been in no way affected by the boiling, is abandoned to the air of the chamber.

With such chambers I tested, in the autumn and winter of 1875-'76, infusions of the most various kinds, embracing natural animal liquids, the flesh and viscera of domestic animals, game, fish, and vegetables. More than fifty moteless chambers, each with its series of infusions, were then tested, many of them repeatedly. There was no shade of uncertainty in any of the results. In every instance we had, within the chamber, perfect limpidity and sweetness, which in some cases lasted for more than a year—without the chamber, with the same infusion, putridity, and its characteristic smells. In no instance was the least countenance lent to the notion that an infusion deprived by heat of its inherent life, and placed in contact with air cleansed of its visibly suspended matter, has any power whatever to generate life anew.

Remembering, then, the number and variety of the infusions employed, and the strictness of our adherence to the rules of preparation laid down by the heterogenists themselves; remembering that we have operated upon the very substances recommended by them as capable of furnishing even in untrained hands easy and decisive proofs of spontaneous generation, and that we have added to their substances many others of our own—if this pretended generative power were a reality, surely it must have manifested itself somewhere. Speaking roundly, I should say that at least 500 chances have been given to it, but it has nowhere appeared. The argument is now to be closed and clinched by an experiment which will remove every residue of doubt as to the ability of the infusions to sustain life. We open the backdoors of our sealed chambers, and permit the common air with its floating particles to have access to our tubes. For three months they have remained pellucid and sweet—flesh, fish, and vegetable extracts, purer than ever cook manufactured. Three days' exposure to the dusty air suffices to render them muddy, fetid, and swarming with infusorial life. The liquids are thus proved, one and all, ready for putrefaction when the contaminating agent is applied. I invite my colleague to reflect on these facts. How will he account for the absolute immunity of a liquid exposed for months in a warm room to optically pure air, and its infallible putrefaction in a few days when exposed to dust-laden air? He must, I submit, bow to the conclusion that the dust-particles are the cause of putrefactive life. And, unless he accepts the hypothesis that these particles, being dead in the air, are, in the liquid, miraculously kindled into living things, he must conclude that the life we have observed springs from germs or organisms diffused through the atmosphere.

The experiments with hermetically-sealed flasks have reached the number of 940. A sample group of 130 of them were laid before the Royal Society on January 13, 1876. They were utterly free from life, having been completely sterilized by three minutes' boiling. I took special care that the temperatures to which the flasks were exposed should include those previously alleged to be efficient. I copied, indeed, accurately the conditions laid down by our most conspicuous heterogenist, but I failed to corroborate him. He then laid stress on the question of warmth, suddenly adding 30° to the temperatures with which both he and I had previously worked. Waiving all argument or protest against the caprice thus manifested, I met this new requirement also. The sealed tubes, which had proved barren in the Royal Institution, were suspended in perforated boxes, and placed under the supervision of an intelligent assistant in the Turkish Bath in Jermyn Street. From two to six days had been allowed for the generation of organisms in hermetically-sealed tubes. Mine remained in the washing-room of the bath for nine days. Thermometers placed in the boxes, and read off twice or three times a day, showed the temperature to vary from a minimum of

101° to a maximum of 112° Fahr. At the end of nine days the infusions were as clear as at the beginning. They were then removed to a warmer position. A temperature of 115° had been mentioned as particularly favorable to spontaneous generation. For fourteen days the temperature of the Turkish bath hovered about this point, falling once as low as 106°, reaching 116° on three occasions, 118° on one, and 119° on two. The result was quite the same as that just recorded. The higher temperatures proved perfectly incompetent to develop life.

Taking the actual experiment we have made as a basis of calculation, if our 940 flasks were opened on the hay-loft of the Bel Alp 858 of them would become filled with organisms. The escape of the remaining 82 strengthens our case against the heterogenists, proving as it does conclusively that not in the air, nor in the infusions, nor in anything continuous diffused through the air, but in *discrete particles* nourished by the infusions, we are to seek the cause of life. Our experiment proves these particles to be in some cases so far apart on the hay-loft as to permit 10 per cent, of our flasks to take in air without contracting contamination. A quarter of a century ago Pasteur proved the cause of "so-called spontaneous generation" to be *discontinuous.* I have already referred to his observation that 12 out of 20 flasks opened on the plains escaped infection, while 19 out of 20 flasks opened on the Mer de Glace escaped. Our own experiment at the Bel Alp is a more emphatic instance of the same kind, 90 per cent, of the flasks opened in the hay-loft being smitten, while not one of those opened on the free mountain-ledge was attacked. The power of the air as regards putrefactive infection is incessantly changing through natural causes, and we are able to alter it at will. Of a number of flasks opened in 1876 in the laboratory of the Royal Institution, 42 per cent, were smitten, while 58 per cent, escaped. In 1877 the proportion in the same laboratory was 68 per cent, smitten to 32 intact. The greater mortality, so to speak, of the infusions in 1877 was due to the presence of hay which diffused its germinal dust in the laboratory air, causing it to approximate, as regards infective virulence, to the air of the Alpine loft. I would ask my friend to bring his scientific penetration to bear upon all the foregoing facts. They do not prove spontaneous generation to be "impossible." My assertions, however, relate not to "possibilities," but to *proofs,* and the experiments just described do most distinctly prove the evidence on which the heterogenist relies to be written on waste paper.

My friend will not, I am persuaded, dispute these results; but he may be disposed to urge that other able and honorable men working at the same subject have arrived at conclusions different from mine. Most freely granted, but let me here recur to the remarks already made in speaking of the experiments of Spallanzani, to the effect that the failure of others to confirm his results by no means upsets their evidence. To fix the ideas, let us suppose that my colleague comes to the laboratory of the Royal Institution, repeats there my experiments, and obtains confirmatory results; and that he then goes to University or King's College, where, operating with the same infusions, he obtains contradictory results. Will he be disposed to conclude that the self-same substance is barren in Albemarle Street and fruitful in Gower Street or the Strand? His Alpine experience has already made known to him the literally infinite differences existing between different samples of air as regards their capacity for putrefactive infection. And, possessing this knowledge, will he not substitute, for the adventurous conclusion that an organic infusion is barren at one place and spontaneously generative at another, the more rational and obvious one that the air of the two localities which has had access to the infusion is infective in different degrees?

As regards workmanship, moreover, he will not fail to bear in mind that *fruitfulness* may be due to errors of manipulation, while *barrenness* involves the presumption of correct experiment. It is only the careful worker that can secure the latter, while it is open to every novice to obtain the former. Barrenness is the result at which the conscientious experimenter, whatever his theoretic convictions may be, ought to aim, omitting no pains to secure it, and resorting, only when there is no escape from it, to the conclusion that the life observed comes from no source which correct experiment could neutralize or avoid. Let us again take a definite case. Supposing my colleague to operate with the same apparent care on 100 infusions—or rather on 100 samples of the same infusion—and that 50 of them prove fruitful and 50 barren. Are we to say that the evidence for and against heterogeny is equally balanced? There are some who would not only say this, but who would treasure up the 50 fruitful flasks, as "positive results, and lower the evidential value of the 50 barren flasks by labeling them "negative" results. This, as shown by Dr. William Roberts, is an exact inversion of the true order of the terms positive and negative. Not such, I trust, would be the course pursued by my friend. As regards the 50 fruitful flasks he would, I doubt not, repeat the experiment with redoubled care and scrutiny, and, not by one repetition only, but by many, assure himself that he had not fallen into error. Such faithful scrutiny fully carried out would infallibly lead him to the conclusion that here, as in all other cases, the evidence in favor of spontaneous generation crumbles in the grasp of the competent inquirer.

The botanist knows that different seeds possess different powers of resistance to heat. Some are killed by a momentary exposure to the boiling temperature, while others withstand it for several hours. Most of our ordinary seeds are rapidly killed, while Pouchet made known to the Paris Academy of Sciences, in 1866, that certain seeds, which had been transported in fleeces of wool from Brazil, germinated after four hours' boiling. The germs of the air vary as much among themselves as the seeds of the botanist. In some localities the diffused germs are so tender that boiling for five minutes, or even less, would be sure to destroy them all; in other localities the diffused germs are so obstinate that many hours' boiling would be requisite to deprive them of their power of germination. The absence or presence of a truss of desiccated hay would produce differences as great as those here described. The greatest endurance that I have ever observed—and I believe it is the greatest on record—was a case of survival after eight hours' boiling. As regards their power of resisting heat, the infusorial germs of our atmosphere might be classified under the following and intermediate heads: Killed in five minutes; not killed in five minutes but killed in fifteen; not killed in fifteen minutes but killed in thirty; not killed in thirty minutes but killed in an hour; not killed in an hour but killed in two hours; not killed in two but killed three hours; not killed in three but killed in four hours. I have had several cases of survival after four and five hours' boiling, some survivals after six, and one after eight hours' boiling. Thus far has experiment actually reached, but there is no valid warrant for fixing upon even eight hours as the extreme limit of vital resistance. Probably more extended researches (though mine have been very extensive) would reveal germs more obstinate still. It is also certain that we might begin earlier, and find germs which are destroyed by a temperature far below that of boiling water. In the presence of such facts, to speak of *a* death-point of bacteria and their germs would be mere nonsense—but of this more anon.

We have now to test one of the principal foundations of the doctrine of spontaneous generation as formulated in this country. With this view, I place before my friend and co-inquirer two liquids which have been kept for six months in one of our sealed chambers, exposed to optically pure air. The one is a mineral solution containing in proper proportions all the substances which enter into the composition of bacteria, the other is an infusion of turnip—it might be any one of a hundred other infusions, animal or vegetable. Both liquids are as clear as distilled water, and there is no trace of life in either of them. They are, in fact, completely sterilized. A mutton-chop, over which a little water has been poured to keep its juices from drying up, has lain for three days upon a plate in our warm room. It smells offensively. Placing a drop of the fetid mutton-juice under a microscope, it is found swarming with the bacteria which live by putrefaction, and without which no putrefaction can occur. With a speck of the swarming liquid I inoculate the clear mineral solution and the clear turnip-infusion, as a surgeon might inoculate an infant with vaccine lymph. In four-

and-twenty hours the transparent liquids have become turbid throughout, and, instead of being barren as at first, they are teeming with life. The experiment may be repeated a thousand times with the same invariable result. To the naked eye the liquids at the beginning were alike, being both equally transparent—to the naked eye they are alike at the end, being both equally muddy. Instead of putrid mutton-juice we might take as a source of infection any one of a hundred other putrid liquids, animal or vegetable. So long as the liquid contains the living bacteria, a speck of it communicated to the clear mineral solution, or to the clear turnip-infusion, produces in twenty-four hours the effect that we have described.

We now vary the experiment thus: Opening the back-door of another closed chamber which has contained for months the pure mineral solution and the pure turnip-infusion side by side, I drop into each of them a small pinch of laboratory dust. The effect here is tardier than when the speck of putrid liquid was employed. In three days, however, after its infection with the dust, the turnip-infusion is muddy, and swarming as before with bacteria. But what about the mineral solution which, in our first experiment, behaved in a manner undistinguishable from the turnip-juice? At the end of three days there is not a bacterium to be found in it. At the end of three weeks it is equally innocent of bacterial life. We may repeat the experiment with the solution and the infusion a hundred times, with the same invariable result. Always in the case of the latter the sowing of the atmospheric dust yields a crop of bacteria—never in the former does the dry germinal matter kindle into active life. What is the inference which the reflecting mind must draw from this experiment? Is it not as clear as day that while both liquids are able to feed the bacteria and to enable them to increase and multiply, *after they have been once fully developed,* only one of the liquids is able to develop into active bacteria the germinal dust of the air?

I invite my friend to reflect upon this conclusion; he will, I think, see that there is no escape from it. He may, if he prefers it, hold the opinion, which I consider erroneous, that bacteria exist in the air, not as germs but as desiccated organisms. The inference remains that, while the one liquid is able to force the passage from the inactive to the active state, the other is not.

But this is not at all the inference which has been drawn from experiments with the mineral solution. Seeing its ability to nourish bacteria when once inoculated with the living active organism, and observing that no bacteria appeared in the solution after long exposure to the air, the inference was drawn that *neither bacteria nor their germs existed in the air.* Throughout Germany the ablest literature of the subject, even that opposed to heterogeny, is infected with this error; while heterogenists at home and abroad have based upon it a triumphant demonstration of their doctrine. It is proved, they say, by the deportment of the mineral solution that neither bacteria nor their germs exist in the air; hence, if, on exposing a thoroughly sterilized turnip-infusion to the air, bacteria appear, they must of necessity have been spontaneously generated. In the words of Dr. Bastian, uttered not in a popular book, but in the "Proceedings of the Royal Society," with reference to this very experiment: "We can only infer that while the boiled saline solution is quite incapable of engendering bacteria, such organisms are able to arise *de novo* in the boiled organic infusion." I would ask my eminent colleague what he thinks of this reasoning now? The *datum* is, "A mineral solution exposed to common air does not develop bacteria:" the *inference* is, "Therefore, if a turnip-infusion similarly exposed develop bacteria, they must be spontaneously generated." The inference, on the face of it, is an unwarranted one. But, while as matter of logic it is inconclusive, as matter of fact it is chimerical. London air is as surely charged with the germs of bacteria as London chimneys are with smoke. The inference just referred to is completely disposed of by the simple question: "Why, when your sterilized organic infusion is exposed to optically pure air, should this generation of life *de novo* utterly cease? Why should I be able to preserve my turnip-juice side by side with your saline solution for the three hundred and sixty-five days of the year, in free connection with the general atmosphere, on the sole condition that the portion of that atmosphere in contact with the juice shall be visibly free from floating dust, while three days' exposure to that dust fills it with bacteria?" Am I over-sanguine in hoping that, as regards the argument here set forth, he who runs may read, and he who reads may understand? Let me add, however, that while exposing the fallacy of the inferences drawn from it, I regard the observation that the boiled saline solution can sustain the developed organisms, while it cannot develop them from the dry germinal matter of the air, as an important addition to our knowledge. We are indebted for it to Dr. Burdon-Sanderson, who soon saw that his first interpretation of it went too far, and who, in a communication recently presented to the Royal Society, abandons the interpretation altogether.

We now proceed to the calm and thorough consideration of another subject, more important if possible than the foregoing one, but like it somewhat difficult to seize by reason of the very opulence of the phraseology, logical and rhetorical, in which it has been set forth. The subject now to be considered relates to what has been called "the death-point of bacteria." Those who happen to be acquainted with the modern English literature of the question will remember how challenge after challenge has been issued to panspermatists in general, and to one or two home workers in particular, to come to close quarters on this cardinal point. It is obviously the stronghold of the English heterogenist. "Water," he says, "is boiling merrily over a fire, when some luckless person upsets the vessel so that the heated fluid exercises its scathing influence upon an uncovered portion of the body—hand, arm, or face. Here at all events there is no room for doubt. Boiling water unquestionably exercises a most pernicious and rapidly-destructive effect upon the living matter of which we are composed." And, lest it should be supposed that it is the high organization which, in this case, renders the body susceptible to heat, he refers to the action of boiling water on the hen's-egg to dissipate the notion. "The conclusion," he says, "would seem to force itself upon us that there is something intrinsically deleterious in the action of boiling water upon living matter—whether this matter be of high or of low organization." Again, at another place, "It has been shown that the briefest exposure to the influence of boiling water is destructive of all living matter." Throughout his prolonged disquisitions on this subject, Dr. Bastian makes special kinds of living matter do duty for *all* kinds. To invalidate the foregoing statements it is only necessary to say that eight years before they were made it had been known to the wool-staplers of Elbœuf, and Pouchet had published the fact in the *Comptes-Rendus of the Paris Academy of Sciences* that the desiccated seeds of the Brazilian plant *medicago* survived fully four hours' boiling. Pouchet himself boiled the seeds, and found some of them swollen and disintegrated, while others remained hard and unswollen. Sown in the same earth, the latter germinated while the former did not. So much for the heterogenist's mistake regarding ordinary seeds; we must now examine whether no error underlies his experiments and his reasonings as to "the death-point of bacteria."

The experiments already recorded plainly show that there is a marked difference between the dry bacterial matter of the air, and the wet, soft, and active bacteria of putrefying organic liquids. The one can be luxuriantly bred in the saline solution, the others refuse to be born there, while both of them are copiously developed in a sterilized turnip-infusion. Inferences, as we have already seen, founded on the deportment of the one liquid cannot with the warrant of scientific logic be extended to the other. But this is exactly what the heterogenist has done, thus repeating, as regards the death-point of bacteria, the error into which he fell concerning the germs of the air. Let us boil our muddy mineral solution with its swarming bacteria for five minutes. In the soft, succulent condition in which they exist in the solution not one of them escapes destruction. The same is true of the turnip-infusion if it be inoculated with the living bacteria only—the aërial dust being carefully excluded. In both cases the dead organisms sink to the bottom of the liquid, and without reinoculation no fresh organisms will arise. But the case is entirely different when we inoculate our turnip-infusion with the desiccated germinal matter afloat in the air.

The "death-point" of bacteria is the maximum temperature at which they can live, or the minimum temperature at which they cease to live. If, for example, they survive a temperature of 140°, and do not survive a temperature of 150°, the death-point lies somewhere between these two temperatures. Vaccine lymph, for example, is proved by Messrs. Braidwood and Vacher to be deprived of its power of infection by brief exposure to a temperature between 140° and 150° Fahr. This may be regarded as the death-point of the lymph, or rather of the particles diffused in the lymph, which constitute the real contagium. If no time, however, be named for the application of the heat, the term "death-point" is a vague one. An infusion, for example, which will resist five hours' continuous exposure to the boiling temperature, will succumb to five days' exposure to a temperature 50° below that of boiling. The fully-developed, soft bacteria of putrefying liquids are not only killed by five minutes' boiling, but by less than a single minute's boiling—indeed, they are slain at about the same temperature as the vaccine. The same is true of the plastic, active bacteria of the turnip-infusion. But, instead of choosing a putrefying liquid for inoculation, let us prepare and employ our inoculating substance in the following simple way: Let a small wisp of hay, desiccated by age, be washed in a glass of water, and let a perfectly sterilized turnip-infusion be inoculated with the washing liquid. After three hours' continuous boiling the infusion thus infected will often develop luxuriant bacterial life. Precisely the same occurs if a turnip infusion be prepared in an atmosphere well charged with desiccated hay-germs. The infusion in this case infects itself without special inoculation, and its subsequent resistance to sterilization is often very great. On the 1st of March last I purposely infected the air of our laboratory with the germinal dust of a sapless kind of hay mown in 1875. Ten groups of flasks were charged with turnip infusion prepared in the infected laboratory, and were afterward subjected to the boiling temperature for periods varying from 15 minutes to 240 minutes. Out of the ten groups only one was sterilized—that, namely, which had been boiled for four hours. Every flask of the nine groups which had been boiled for 15, 30, 45, 60, 75, 90, 105, 120, and 180 minutes respectively, bred organisms afterward. The same is true of other vegetable infusions. On the 28th of February last, for example, I boiled six flasks, containing cucumber-infusion prepared in an infected atmosphere, for periods of 15, 30, 45, 60, 120, and 180 minutes. Every flask of the group subsequently developed organisms. On the same day, in the case of three flasks, the boiling was prolonged to 240, 300, and 360 minutes; and these three flasks were completely sterilized. Animal infusions, which under ordinary circumstances are rendered infallibly barren by five minutes' boiling, behave like the vegetable infusions in an infective atmosphere. On the 30th of March, for example, five flasks were charged with a clear infusion of beef and boiled for 60 minutes, 120 minutes, 180 minutes, 240 minutes, and 300 minutes respectively. Every one of them became subsequently crowded with organisms, and the same happened to a perfectly pellucid mutton-infusion prepared at the same time. The cases are to be numbered by hundreds in which similar powers of resistance were manifested by infusions of the most diverse kinds.

In the presence of such facts I would ask my eminent colleague whether it is necessary to dwell for a single instant on the one-sidedness of the evidence which led to the conclusion that all living matter has its life destroyed by "the briefest exposure to the influence of boiling water." An infusion proved to be barren by six months' exposure to moteless air kept at a temperature of 90° Fahr., when inoculated with full-grown, active bacteria, fills itself in two days with organisms so sensitive as to be killed by a few minutes' exposure to a temperature much below that of boiling water. But the extension of this result to the desiccated germinal matter of the air is without warrant or justification. This is obvious without going beyond the argument itself. But we have gone far beyond the argument and proved by multiplied experiment the alleged destruction of all living matter by the briefest exposure to the influence of boiling water to be a delusion. The whole logical edifice raised upon this basis falls, therefore, to the ground; and the argument that bacteria and their germs being destroyed at 140° must, if they appear after exposure to 212°, be spontaneously generated, is, I trust, silenced forever.

Through the precautions, variations, and repetitions observed and executed with the view of rendering its results secure, the separate vessels employed in this inquiry have mounted up in two years to nearly 10,000. Here, however, and with good reason, the editor cries, "Halt!" I had hoped, when I began, to carry the argument further. Besides the philosophic interest attaching to the problem of life's origin, which will be always immense, there are the practical interests involved in the application of the doctrines here discussed to surgery and medicine. The antiseptic system, at which I have already glanced, illustrates the manner in which beneficent results of the gravest moment follow in the wake of clear theoretic insight. Surgery was once a noble art; it is now, as well, a noble science. Prior to the introduction of the antiseptic system, the thoughtful surgeon could not have failed to learn empirically that there is something in the air which often defeated the most consummate operative skill. That something the antiseptic treatment destroys or renders innocuous. At King's College Mr. Lister operates and dresses while a fine shower of mixed carbolic acid and water, produced in the simplest manner, falls upon the wound, the lint and gauze employed in the subsequent dressing being duly saturated with the antiseptic. At St. Bartholomew's Mr. Callender employs the dilute carbolic acid without the spray; but, as regards the real point aimed at—the preventing of the wound from becoming a nidus for the propagation of septic bacteria—the practice in both hospitals is the same. Commending itself as it does to the scientifically-trained mind, the antiseptic system has struck deep root in Germany.

It would also have given me pleasure to point out the present position of the "germ-theory" in reference to the phenomena of infectious disease, distinguishing arguments based on analogy—which, however, are terribly strong—from those based on actual observation. I should have liked to follow up the account I have already given of the truly excellent researches of a young and an unknown German physician named Koch, on splenic fever, by an account of what Pasteur has recently done with reference to the same subject. Here we have before us a living contagium of the most fatal power, which we can follow from the beginning to the end of its life-cycle. We find it in the blood or spleen of a smitten animal in the state say of short motionless rods. We place these rods in a nutritive liquid on the warm stage of the microscope, and see them lengthening into filaments which lie side by side, or, crossing each other, become coiled into knots of a complexity not to be unraveled. We finally see those filaments resolving themselves into innumerable spores, each with death potentially housed within it, yet not to be distinguished microscopically from the harmless germs of *Bacillus subtilis.* The bacterium of splenic fever is called *Bacillus anthracis.* This formidable organism was shown to me by M. Pasteur in Paris last July. His recent investigations regarding the part it plays pathologically certainly rank among the most remarkable labors of that remarkable man. Observer after observer had strayed and fallen in this land of pitfalls, a multitude of opposing conclusions and mutually-destructive theories being the result. In association with his younger physiological colleague M. Joubert, Pasteur struck in amid the chaos, and soon reduced the whole of it to harmony. They proved, among other things, that in cases where previous observers in France had supposed themselves to be dealing solely with splenic fever, another equally virulent factor was simultaneously active. Splenic fever was often overmastered by septicæmia, and results due solely to the latter had been frequently made the ground of pathological inferences regarding the character and cause of the former. Combining duly the two factors, all the previous irregularities disappeared, every result obtained receiving the fullest explanation. On studying the account of this masterly investigation, the words wherewith Pasteur himself feelingly alludes to the difficulties and dangers of the experimenter's art came home to me with especial force: "*J'ai tant de fois éprouve que dans cet art difficile de l'expérimentation les plus hahiles bronchent à chaque pas, et que l'interprétation des faits n'est pas moins périlleuse.*"

www.ingramcontent.com/pod-product-compliance
Lightning Source LLC
Chambersburg PA
CBHW030914180526
45163CB00004B/1823